DODD, MEAD WONDERS BOOKS include WONDERS OF:

ALLIGATORS AND CROCODILES.
    Blassingame
ANIMAL NURSERIES. Berrill
BARNACLES. Ross and Emerson
BAT WORLD. Lavine
BEYOND THE SOLAR SYSTEM.
    Feravolo
BISON WORLD. Lavine and Scuro
CACTUS WORLD. Lavine
CAMELS. Lavine
CARIBOU. Rearden
CATTLE. Scuro
CORALS AND CORAL REEFS.
    Jacobson and Franz
CROWS. Blassingame
DINOSAUR WORLD. Matthews
DONKEYS. Lavine and Scuro
DUST. McFall
EAGLE WORLD. Lavine
ELEPHANTS. Lavine and Scuro
FLY WORLD. Lavine
FROGS AND TOADS. Blassingame
GEESE AND SWANS. Fegely
GEMS. Pearl
GOATS. Lavine and Scuro
GRAVITY. Feravolo
HAWK WORLD. Lavine
HERBS. Lavine
HOW ANIMALS LEARN. Berrill
HUMMINGBIRDS. Simon
JELLYFISH. Jacobson and Franz
KELP FOREST. Brown
LLAMAS. Perry
LIONS. Schaller
MARSUPIALS. Lavine

MEASUREMENT. Lieberg
MICE. Lavine
MONKEY WORLD. Berrill
MOSQUITO WORLD. Ault
OWL WORLD. Lavine
PELICAN WORLD. Cook and Schreiber
PONIES. Lavine and Casey
PRAIRIE DOGS. Chace
PRONGHORN. Chace
RACCOONS. Blassingame
RATTLESNAKES. Chace
ROCKS AND MINERALS. Pearl
SEA GULLS. Schreiber
SEA HORSES. Brown
SEALS AND SEA LIONS. Brown
SNAILS AND SLUGS. Jacobson and
    Franz
SPIDER WORLD. Lavine
SPONGES. Jacobson and Pang
STARFISH. Jacobson and Emerson
STORKS. Kahl
TERNS. Schreiber
TERRARIUMS. Lavine
TREE WORLD. Cosgrove
TURTLE WORLD. Blassingame
WILD DUCKS. Fegely
WOODS AND DESERT AT NIGHT. Berrill
WORLD OF THE ALBATROSS. Fisher
WORLD OF BEARS. Bailey
WORLD OF HORSES. Lavine and Casey
WORLD OF SHELLS. Jacobson and
    Emerson
WORLD OF WOLVES. Berrill
YOUR SENSES. Cosgrove

# Wonders of Cattle
## Vincent Scuro

ILLUSTRATED WITH
PHOTOGRAPHS AND OLD PRINTS

DODD, MEAD & COMPANY · NEW YORK

To Sigmund A. Lavine

ILLUSTRATIONS COURTESY OF: American Angus Association, 42, 54; American Brahman Breeders Association, 10–11, 45; American Guernsey Cattle Club, 57; American Milking Shorthorn Society, 48 *top*; American National Cattlemen's Association, 33, 51, 58, 66; American Red Poll Association, 48 *bottom*; American Shorthorn Association, 34, 44; *Brangus Journal*, San Antonio, Texas, photo by Caren Cowan, 47; The Brown Swiss Cattle Breeder's Association, 40; Calgary Stampede and Exhibition, 74; Cheyenne, Wyoming, Frontier Days, photo by Randall A. Wagner, 71, 72; Colorado Historical Society, 27, 29, 31, 32; Samantha Cozi, 17, 36; Department of Agriculture and Fisheries, States of Jersey, Channel Islands, 39 *top*, 76, 77 *top*; Holstein Association of America, 2, 6, 37, 56, 77 *bottom*; Tina Krettecos, 55; Merrill Lynch and Co., Inc., 22; National Archives, 8, 13, 28, 59; *Polled Hereford World*, Kansas City, Missouri, 41; Red Angus Association of America, 43; Santa Gertrudis Breeders International, Kingsville, Texas, 46; Vincent Scuro, 21, 23, 65, 69, 75; Spanish National Tourist Office, 15; States of Jersey Tourism Committee, 39 *bottom*; United Nations, 12, 19; USDA Photo, 26, 52, 53, 60, 61, 62; Vermont Travel Division, 10 *top*.

1   2   3   4   5   6   7   8   9   10

Library of Congress Cataloging in Publication Data

Scuro, Vincent.
    Wonders of cattle.

    Includes index.   *a9S*
    SUMMARY: Traces the ancestry of cattle, their place
in literature and lore, their history in the United
States, and their characteristics, habits, and useful-
ness.
    1. Cattle—Juvenile literature. [1. Cattle]
I. Title.
SF197.S38        636.2        80–1019
ISBN 0–396–07892–3

# Contents

1   Introducing Cattle    7
    *Early Ancestors*    10
    *Domestication*    11
2   Cattle Tales    14
3   Cattle and the Old West    24
4   Some of the Breeds    35
    *Dairy Cattle Breeds*    36
    *Beef Cattle Breeds*    41
    *Dual-purpose Cattle Breeds*    47
5   The Cattle Industry    50
    *Breeding*    50
    *Dairying*    54
    *Ranching*    58
6   A Very Valuable Animal    64
    *Milk*    64
    *Meat*    65
    *Leather*    68
    *Other Uses*    68
7   Rodeos and Shows    70
    Index    78

*"The friendly cow, all black and white . . ."*

# 1 Introducing Cattle

Of all the earth's creatures, none are as useful to man's well-being as cattle. Not only do these animals provide meat and milk for nourishment but they are also an important source of leather. In many countries, cattle are still the principal beasts of burden. Furthermore, they are as abundant as they are valuable. Approximately 1.2 billion are distributed around the globe. In some areas, there are more cattle per square mile than there are people. It is no wonder that cattle supply half of the world's meat, 95 percent of the milk, and 80 percent of the leather.

Cattle were equally valuable to ancient peoples. The worth of property was often measured in terms of cattle. Like money, they could be used to purchase land, services, or other livestock. Some cultures would not allow a marriage to take place until the father of the bride gave several of these animals to his prospective son-in-law.

Lexicographers—compilers of dictionaries—are not certain of the origin of the word cattle. It may have come from the Middle English noun *catel*, which was used to describe personal property. Fourteenth-century writers wrote *cattell* when referring to animals held as livestock, and *chattel* for any movable form of property. Eventually, the present form emerged.

*Farmer using oxen to pull his plow in Alabama, about 1920*

While some books speak of wild cattle, this is technically incorrect. True cattle must be domesticated. Similarly, common speech tends to use cows and cattle interchangeably. Actually, cows are female cattle. In fact, a female is referred to as a heifer until she has given birth, and thereafter is called a cow.

All young cattle are known as calves until they are a year old, at which time they become yearlings. A male, however, may be called a bull at any age. When males are raised solely for beef they are castrated and classified as steers. An ox is a male used as a beast of burden. In Europe, the plural of ox—oxen—is often used interchangeably with cattle. Incidentally, when speaking or writing about a specific number of these animals, they should be referred to as head of cattle. American cattle raisers take great pride in describing their ranches in terms of the number of head of cattle they have, not the acreage of the land.

Zoologists—scientists who study animals—have placed cattle in the Bovidae family, a large group which also includes bison, sheep, goats, and buffalo. Like other members of the Bovidae family, cattle are ruminants, or cud-chewers. Every ruminant has a stomach with four chambers. When food is swallowed, it travels to the first two chambers, where it is attacked by digestive enzymes. These enzymes break down the cellulose the food contains. The cud, as it is called, is then returned to the mouth, where it is rechewed. Once the cud has been thoroughly masticated, it is again swallowed and passed into the second two stomach chambers where it is reduced to a pulp that can be digested easily. This process enables cattle and other ruminants to eat a large amount of food quickly, with digestion taking place later.

To distinguish them from other ruminants, cattle have been placed in the Bovinae subfamily. Thus, they are often referred to as bovines. Additionally, cattle belong to a subdivision of Bovinae—the *Bos* genus—which is, in turn, divided into smaller groups called species. There are two species of cattle. *Bos taurus* (European cattle) is the predominant type found in Europe, the Western Hemisphere, and Oceania. Members of *Bos indicus* (Indian cattle) live primarily in India, Asia, the Middle East, and parts of Africa. *Indicus* is often called the zebu. In recent times, zebus have been imported to the United States, Canada, and Australia for crossbreeding with *taurus* types.

In some respects, European and Indian cattle are similar. A bulky body, straight spine, and broad, flat teeth are common features. All cattle have four cloven hoofs, each hoof with two toes. Breeds of either species may have short horns, long horns, or no horns—the latter are referred to as polled cattle. However there are some differences. *Bos indicus* has a large fatty hump

*A herd of Holsteins crossing a Vermont road*—Bos taurus

on its back and shoulders, and huge folds of skin (called dewlaps) under its chin. These characteristics are usually absent on breeds of *Bos taurus*.

### Early Ancestors

*Eotragus*, an antelope-like creature that lived about 25 million years ago, is thought to be the progenitor of all members of the Bovidae family. The immediate ancestor of *Bos taurus* is believed to be *Bos primigenius*, a virtually extinct species known

as the aurochs. *Bos indicus* may have evolved from a subspecies of aurochs—*Bos primigenius nomadicus*, which also may be extinct.

Fossil discoveries have revealed that the aurochs, a long-horned beast, originated in northern India over a million years ago. During the Pleistocene period (between one million and ten thousand years ago), the aurochs migrated to Europe, northern Africa and the rest of Asia.

It is not certain where *Bos longifrons* belongs in the ancestry of cattle. Was it a distinct wild species or a domesticated version of *primigenius*? No one knows for sure. This confusion is probably caused by the fact that it is impossible to distinguish the remains of wild animals captured by early man and brought back to his camps from those of the domesticated creatures that lived with him. Incidentally, the fossils of *longifrons* that have been unearthed in Ireland show that these creatures had horns that were not very long. Thus textbooks often refer to *longifrons* as "Celtic shorthorns."

## Domestication

Prehistoric bovines were hunted for their meat and hides. Once it was learned that milk could be obtained from them too, these creatures were captured and domesticated.

*Brahmans on the Runnells-Pierce Ranch in Bay City, Texas*—Bos indicus

A complete discussion of how cattle were domesticated is beyond the scope of a book this size. However, it is worth noting that the social nature of prehistoric bovines and their ability to subsist on a variety of foods made domestication relatively easy.

While the exact date is not known, it is generally accepted that cattle were first domesticated about 6000 B.C. in the region that is now called Iraq. As people moved from one area of the world to another, they brought their cattle with them. It was soon discovered that two oxen (called a yoke) could haul a wagon loaded with heavy goods and that one could pull a plow all day in intense heat without tiring. The ancient Egyptians

*A typical ox-cart used for transportation in Ceylon*

*Oxen long have been used in logging operations such as this one near Smithfield, North Carolina.*

used these beasts to tread out grain on their threshing floors. Their drawings dating from the third century B.C. show longhorn, shorthorn, and polled cattle. By 200 B.C., many cultures were using ox-driven water wheels for irrigation.

From 27 B.C. to about A.D. 395, the Romans distributed cattle throughout their empire. Some of the modern English breeds are believed to be descendants of Celtic shorthorns that mixed with domesticated stock brought to the British Isles from Europe by Caesar's legions. As time passed, these animals were transported to the Western Hemisphere and Australia.

# 2 Cattle Tales

There is no way of knowing if Stone Age man endowed cattle with magical powers. However, it can be assumed that bovines were feared. The lifelike paintings discovered in the caves of Spain and France depict wild bulls being stabbed with darts and spears. Perhaps these drawings were made in underground caverns because the artists did not want to be discovered by the beasts they sought to destroy. Stone Age cults also drew a relationship between the changing phases of the moon and the crescent-shaped horns of the ox. During rituals for the moon goddess, Stone Age man slaughtered cows. Certain organs were burned during the ceremony. The rest were cooked, then eaten.

Many myths of Crete (the Minoan civilization) are based on sacrifices involving cattle. The most famous tells of a monster, called the Minotaur, which had the torso of a man and the head of a bull. The Minotaur lived in a huge maze called the Labyrinth. Every nine years, King Minos of Crete demanded that the people of Athens send seven boys and an equal number of girls to be sacrificed to the bull-monster as a tribute. According to legend, an Athenian boy named Theseus was sent to Crete. Refusing to accept his fate, Theseus slew the Minotaur, saving its captives from a terrible death.

In the early part of the twentieth century, archaeologists

14

*Bull fights are a tradition in Spain and Latin America.*

uncovered a labyrinth on the island of Crete. They found something else—a wall painting showing a boy doing a somersault over a bull with a girl trying to catch him. Historians have determined that the Minoans trained performers for a bull-vaulting ceremony to honor Zeus, king of the gods. During this ritual, male and female athletes entered a ring unarmed, waiting for a bull to charge. At the last possible moment, the athletes would try to grasp the bull's horns. If they were successful, they would be tossed in the air when the animal snapped its head upward. With every toss, the spectators believed that their athletes were absorbing the bull's strength. If the athletes failed to grasp the horns, they were severely gored. In another ceremony, several Cretans would slay a charging bull, hoping to obtain its strength by killing it.

Today, the ritual of cattle sacrifice takes place in the bullfighting ring, where the bull always loses. A tradition in Spain and Latin America, this spectacle involves a bull that is first tired out and tormented by mounted lancers (the picadors), and then killed by a matador who uses a sword and a cape. In Portugal, the *salteadores*, who pole vault over charging bulls, are a variation of the Cretan somersaulters.

Bull baiting, a "sport" in which dogs attack a shackled bull and spectators bet on the outcome, has been practiced in all parts of the world. The Baltic Finns had a different reason for killing bovines. They believed that if a cow was not slaughtered at the funeral of its owner, it would die anyway. Cattle occasionally kill humans at the annual "running of the bulls" in Pamplona, Spain, where people try to outsprint bulls that have been let loose in the city's streets before the bullfights. Most participants say they enjoy the danger. Nevertheless, several runners are gored or trampled to death every year.

Cattle as religious symbols throughout the ancient world may be traced to the Egyptian portrayal of the heavens in the form

*Homer's* Odyssey *tells of the sacred white oxen of Thrinacia, which belonged to the sun god.*

of a large cow. The ancient Egyptians had many bovine deities. Nekhbet was called "the great wild cow." Isis was often represented as having the horns of a cow on a human head. Sothis-Sirius was pictured as a cow reclining in a great ship, symbolizing her rule over the heavens. The bull-god, Apis, reigned over the moonlight. A pillar built in honor of Osiris shows the interlocked horns of a cow and a ram holding up the sun. Some Egyptians believed that nailing the skull of a cow over a doorway would ward off evil spirits. Another culture of the ancient Near East, that of the Mesopotamians, believed that their kings were related to bull gods.

The myths of ancient Greece abound with tales of cattle. Homer's *Odyssey* tells of the sacred white oxen of Thrinacia, which belonged to the sun god. In another story, Jason, while seeking the Golden Fleece, meets bulls "wreathed in smoke and blowing fire from their mouths and nostrils." According to legend, the Oracle of Delphi told Kadmos, a Greek boy, to follow a heifer and to build the city of Thebes at the first place it stopped to rest.

Meanwhile, Celtic tribesmen were attributing virility and divine powers to cattle. The cow plays an important role in the story of Bres, a miserly Celt who claimed that the milk of all brown, hairless cows belonged to him. Since these beasts were so few in number, Bres caused many to pass through "a fire of bracken" so that they might become hairless and brown. The Bres tale is probably related to the Celtic ritual of passing cattle through a fire on May Day.

Another Celtic myth tells of a great war in Ireland that was caused by two cows. The bloody history of still another war that was fought over possession of a magical brown bull is told in the famous *Cattle Raid of Cooley*. The Gaelic Celts had a deity, Tuatha Dé Danann, that they believed had the power to control cattle.

The Finns believed in the Maahiset—"those living under the earth"—who were thought to own enormous cattle. A person could obtain one of these great beasts by throwing metal upon its back. Gufittar of the Scandinavian Lapps, an underground dwarf, is credited with hanging a bell around a cow's neck. To this day, cowbells are used to let herdsmen know where their animals are grazing.

Cattle play an important role in the lore of other cultures. The Siberians believed in "Cow-footed Man," an evil spirit who dressed in a peasant costume from the waist up. Below the waist, he had hairy legs that ended in hoofs. In a Mongolian tale, the constellation known to us as Orion is said to have been "born of a cow."

18

*The Hindus, more than any other people, revere cattle.*

The Hindus probably revere cattle more than any other people. The milk, curd, butter, dung, and urine of cows are regarded as having purifying properties. Tanners and those who eat beef are considered to be unclean. In India, the penalty for killing a bovine was once severe—the slayer had to wear the skin of the dead animal for several days while living with the herd!

No people have lived closer to their cattle than the inhabitants of a region referred to as the East African Cattle Area. A number of political, legal, and mystical ideas stem from the dependence of the natives on their bovine herds. The Banya consider cattle second only to man in the order of creation. The Tutsi have songs to praise cows, songs to show that a person should own cows, songs to bring cows home in the evening, songs to take cows to water, and songs to emphasize the importance of cows in tribal history. The Masai believe that drinking the blood of their cattle will give their tribe strength; thus they bleed their bovines for this purpose.

There are numerous references to cattle in the Bible. According to the Book of Exodus, the Lord brought a plague down upon the cattle of the Egyptians because their owners refused to release the Israelites from bondage. The best-known story in the Old Testament is probably that of the Golden Calf, an idol fashioned by the newly released Israelites and later destroyed by Moses, who was angry at his people for worshiping a false god. The "cattle of a thousand hills" (Psalms 50:10) probably refers to the many breeds.

It would take an enormous barn to shelter all of the cattle that have appeared in literary works. One of the earliest Gaelic writings is the ancient *Book of the Dun Cow*. Cossack literature includes "The Straw Ox," one of *Bain's Fairy Tales*. A bull appears in "Who Killed Cock Robin?" But perhaps the best-known bovine of literature and folklore is Babe, the Blue Ox, who appears in the stories of American lumberjack Paul Bunyan. Legend says this ox was so big that if a man stood by Babe's head he "had to carry a telescope to see what the beast's hind legs were doing."

Cattle have plodded through poetry and song for a long time. Children sing of "The Cat and the Fiddle" and the cow that "jumped over the moon." "The Purple Cow" by Gelett Burgess is a well-known rhyme. E. M. Root's "The Cow" praises "God's jolly cafeteria." Edward Lear illustrated his own rhymes about cattle.

Throughout history, artists have been inspired by cattle. A well-known representation is the *Augsburg Aurochs Picture*, painted by an unknown European artist in the sixteenth century. Oxen are commonly found in scenes of the Nativity. The list of artists who have included cattle in their works ranges from those of the Renaissance including Botticelli, Rembrandt, and della Francesca to Americans such as Frederick Remington and Charles M. Russell.

Near Eastern sculptures show bulls being defeated by lions, perhaps representing a victory of cunning and guile over brute strength. Meanwhile, our common speech has been enriched by many sayings featuring cattle. "Milk the cow which is near" reminds us to be happy with what we have, as does "The cow knows not what her tail is worth until she has lost it." Space does not permit the inclusion of every cattle maxim. Besides, listing them all would take "till the cows come home," which is a very long time.

*Taurus, the astrological sign, is represented as a bull with horns.*

Astrologers call the second sign of the Zodiac by the name Taurus, which corresponds to the bull-shaped constellation located between Aries and Gemini. The English language contains numerous idiomatic phrases concerning cattle. A person who attacks a problem fearlessly is said "to take the bull by the horns." Someone in excellent health is considered to be "as strong as an ox." An expletive of amazement is "Holy cow!" One who is thought to be obstinate or stubborn is called "bull-headed."

*The bull is the symbol of Merrill Lynch and Company, a stock brokerage firm.*

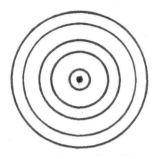

*A target such as this is called a "bull's eye."*

Industry has the cowcatcher, a triangular frame on the front of a locomotive designed for clearing obstructions from the tracks. A bullhorn is a portable loudspeaker. In the financial world, brisk trading of stocks at high exchange rates is said to produce a "bullish" market.

The sports world has its share of cattle terminology. Not only are baseball gloves made from cowhide, but relief pitchers warm up in an area known as the bullpen. Marksmen take aim at a bull's-eye, a circular black spot at the center of a target, and fans of all sports are reminded of the Old West when teams from the University of Texas—the Longhorns—take the field.

# 3 Cattle and the Old West

The immigrants who came to the New World brought a wealth of cattle folklore with them. This collection of proverbs, maxims, and fables has been enriched with stories of the pioneers who tamed the American frontier. Western historian Harry Sinclair Drago has said that cattle "brought romance and color to this era." Most of the tales of cattle and the Old West have achieved legendary stature.

Whether or not the Norsemen took cattle with them on their journeys to North America is a subject which has been debated for years. In any case, it is Christopher Columbus who traditionally is credited with bringing the first bovines to the Western Hemisphere. He did so on his second voyage in 1493. By 1521, large herds of beef cattle had been established in Mexico. When Francisco Vasquez de Coronado started his expedition to search for the mythical "seven cities of Cibola" in 1540, his livestock inventory included five hundred head of Moorish ("black Spanish") and Castilian ("brown Spanish") cattle. These beasts are believed to be the first of their species to enter what is today called the United States.

Most of Coronado's cattle were slaughtered for beef to feed his troops. Some, however, escaped to form feral (domestic stock gone wild) herds. A royal decree had kept these two strains

*A scene on the Plains, print from an old textbook*

separate in Spain, but they ran together freely and mated in the region that is now called Texas. This union of Moorish and Castilian cattle produced offspring with the black coloring of the former and the long, heavy horns of the latter. Eventually, the new strain became known as the Texas Longhorn.

No animal developed by man could have been better suited for the rugged, often inhospitable environment of the American West. A "tall, bony, coarse-headed, coarse-haired, flat-sided, thin-flanked, narrow-hipped critter" with "steel hoofs," the Longhorn could "cross the widest deserts, climb the highest mountains, fight off the fiercest bands of wolves."

Stories are told of steers that weighed from 1000 to 1600 pounds by the time they reached ten years of age—with a span of five to eight feet between the horns! Some had a yellow stripe that ran from the tip of the nose down the spine to a tufted tail that was so long it dragged on the ground.

When Stephen F. Austin and a group of settlers arrived in Texas in 1821, there were feral Longhorns all over the eastern part of the territory. Gradually, these animals were rounded up, forming the basis of the Texas ranch industry. Eventually, the Lone Star State became known as the "Cattle Kingdom." By the

*These Longhorns, part of a government herd in the Wichita Mountain Wildlife Refuge at Lawton, Oklahoma, are the descendants of the great herds that gave fame and fortune to the trail riders of the Old West.*

time daring individuals were seeking gold in California (1849), the demand for beef throughout the United States had increased dramatically.

Longhorns are probably most famous for their ability to cross great distances without tiring. Indeed, this stamina was necessary. The journey from the open range to the market averaged between 1200 and 1500 miles. That they were forced to make this trip (called a trail drive) in six to eight weeks has added to the Longhorn legend.

Harry Drago has written that cattle raising "had an impact on the frontier economy that was second only to mining." It also gave a new name to an old occupation. The herdsman became a cowboy.

Originally, a cowboy was a Tory soldier who, during the

American Revolution, hid in a thicket and jingled a cowbell to lure a Colonial soldier into musket range. It wasn't until 1830 that "cowboy" came to mean a man who rounded up and herded cattle from the back of a horse. In Mexico, he was called a *vaquero*, which is derived from the Spanish word *vaca*—cow.

The life of the nineteenth-century cowboy has been glamorized by motion pictures and television. In reality, it was a rugged existence. Cowboys spent most of their time watching over a herd of cattle, or separating calves from their mothers for branding.

The practice of branding, which dates back to ancient Greece, involves burning a mark (called a brand) on the hair and skin of an animal. Many ranchers have used brands as a means of distinguishing the ownership of cattle, particularly in areas where herds mix together on unfenced land.

In the Old West, branding involved dragging a struggling calf over to a fire where a hot iron was used to burn the hair on

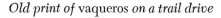

*Old print of* vaqueros *on a trail drive*

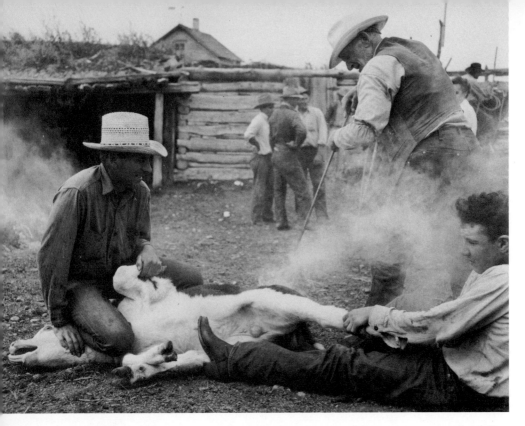

*Cowboys branding a young bull in Montana in the early part of the twentieth century*

its rump in a specific pattern. Each ranch had its own distinctive brand. Cattle were also marked to discourage rustlers—men who made their living by stealing cattle and selling them to unsuspecting buyers. To avoid prosecution, rustlers altered brands by reburning or clipping the hair near the original brand. Eventually, ranchers resorted to cutting calves' ears in a specific pattern—a process called earmarking. These marks were much more difficult for rustlers to change.

Although his fellows ranchers were branding their herds, a Texan named Samuel Maverick (1830–70) refused to do so. As a result, all unbranded calves became known as "mavericks." During this era, a maverick became the property of the first person who branded it. Today, the term is used to describe an individual who takes an independent stand.

During the War Between the States (1861–65), many Texans abandoned their ranches to fight for the Confederacy. The Longhorns, left unattended, continued to breed and multiply. On returning home after the war, the Texans found thousands of Longhorns waiting to be rounded up and taken to market. It is estimated that ten million Longhorns were driven out of Texas in the twenty years after the American Civil War. Most were sold for beef to feed the increasing populations of the northern cities. This demand was reflected in the price of beef. A steer purchased in Texas for four dollars could be sold at a northern market for ten times as much!

Most trail drives were started when several ranchers in a specific area contracted with brokers to sell their cattle for them. The average-size herd in a drive numbered about one thousand animals. The broker assembled the herd and hired a

*Nineteenth-century photograph shows cattle being herded across Milk River during a trail drive.*

trail boss to take charge. In turn, the trail boss hired cowboys to drive the herd to a railhead. From there, the steers would be taken by train to a slaughterhouse in one of the major cities, such as Chicago or St. Louis.

Cowboys on the trail were known as "drovers" because they drove the cattle. Only the best cowboys—those who could withstand the hard work and long hours—were hired. A novice drover (called a wrangler) was brought along to help the cook and take care of the spare horses. A kitchen on wheels called a chuck wagon also traveled with them.

A typical day on a trail drive started at dawn with a quick breakfast of sourdough bread or biscuits. After saddling the horses, the cowboys allowed the herd to graze for awhile before running them at top speed for about fifteen miles. At noon, they stopped for lunch. Then the herd was started on the trail again. Occasionally, one drover would ride off to track down dogies— calves that had wandered away from the herd—and bring them back. After the animals had covered another fifteen miles, they would be bedded down.

At night, drovers worked four-hour shifts, watching over the cattle by riding slowly around them. Some drovers sang or hummed to the herd, believing that a quiet tune would prevent a stampede.

On the trail, cowboys dealt with drought, rustlers, blizzards, and wild animals. Yet the stampede was probably their greatest fear, especially if it happened at night. A flash of lightning or a loud noise could send a herd off and running at full speed. The common belief is that stampeding cattle ran through camp, stomping on sleeping cowboys. Western artist Charles M. Russell would probably disagree. Stampeding cattle were noisy, he said, but they rarely hurt anyone but themselves.

To stop a stampede, cowboys rode around the cattle, trying to get them to run in a circle. Gradually, the circle would be closed,

30

*Loading cattle at Abilene, Kansas, railroad yard about 1868*

thus slowing down the animals. Since a typical herd stretched for about two miles, this could take some time. More often than not, however, a herd would scatter in all directions. Then it took days to round up the critters—if they could be found at all.

Some of the important routes used for the early cattle drives were the Western Trail (southern Texas to Dodge City, Kansas) and the famous Chisholm Trail, which started in San Antonio, Texas, and ran to Abilene, Kansas. The communities at the end of trails were known as cowtowns.

Cowtowns were established so that brokers could bring in the stock and meet with buyers to determine the price per head. If the cattle were thin from the long journey, they could be fattened up before being sold. If the price per head was not as much as the broker wanted, the herd could be grazed outside of town until the price went up. Cowtowns also gave cowboys a place to celebrate the end of a long, hard drive and spend their

*Stock feeding in Austin, Colorado, during the nineteenth century. These are cattle of the type herded during the great trail drives.*

wages on liquor and entertainment.

By the 1870's, cattle ranching had spread throughout the West. With the destruction of the great herds of bison (the buffalo of common speech) and the removal of the Indians from their land, vast tracts of territory became available.

There were no fences on the early ranges. Cattlemen often shared the land. However, "range rights," generally marked by streams, were established between ranches. Cowboys patrolled to keep the cattle within the specified boundaries. When sheep raisers came to the West, there was considerable conflict with cattlemen over these rights. The cattle ranchers charged that sheep were eating all of the vegetation.

At the same time, homesteaders were staking their claims to what they considered to be open rangeland. Many of these settlers came to the West in wagons drawn by cattle. Not only could these beasts of burden provide milk but their cloven hoofs gave them excellent traction through mud or over rocky terrain.

A man named Joseph F. Glidden changed the cattle industry considerably with an invention called barbed wire. These inexpensive strands of wire had small pieces of sharp wire twisted around them at short intervals and were effective in confining stock. Barbed wire enabled cattle ranchers, sheepmen, and farmers to live side-by-side with less conflict.

The invention of refrigeration during the late nineteenth century also changed the cattle industry. Beef could be shipped

*Fence mending is an important ranch chore.*

*Breeds such as the Shorthorn have replaced the Longhorn on American ranches of today.*

by refrigerated railroad car directly to major cities, thus eliminating the need for long trail drives.

An animal with the Longhorn's stamina was no longer needed. New breeds that produce more beef and require less rangeland were introduced to the west in the late 1800's. Today, the Longhorn has joined the American bison on reservations, wildlife preserves, private ranches—and in Western legend.

34

# 4 Some of the Breeds

The development of the 250 modern-day cattle breeds has taken place over hundreds of years. Some of them were developed when farmers in a particular area selected cattle of a type they considered best suited for their location and needs. They mated the animals until offspring with the desired characteristics were produced. Others have been developed by crossing two or more breeds over several generations.

Purebred cattle have the characteristics of a certain breed and a documented ancestry. Registered purebreds have had their lineage officially recorded with a breed association. Grade cattle have the characteristics of a particular breed but are ineligible for registration because one of the parents was not a purebred.

The internal anatomy of dairy and beef cattle is identical. In fact, if one were to examine the skeletons of a beef steer and a dairy cow, it would be impossible to tell them apart. The main difference between dairy and beef cattle is in the development of the udder in females and the amount of flesh on males. Dairy bulls are not as fleshy as their beef counterparts, while the udder of a dairy cow is usually a great deal larger than that of a female bred for beef.

*A herd of grade dairy cows on an Ohio farm*

### Dairy Cattle Breeds

At one time, a cow could supply only enough milk to feed her calf. The milk-producing ability of dairy cattle has increased tremendously during the past century. Today, a single cow can feed twenty calves and also provide milk for human consumption.

Some physical characteristics are common to all dairy cattle, regardless of breed. The typical dairy animal has either an angular or a wedge-shaped body, a long, thin neck, and lean shoulders. As noted, a good milk producer generally has a large udder.

The majority of the dairy cattle in the United States are pure-breds or grades of six breeds—Holstein, Jersey, Guernsey, Brown Swiss, Ayrshire, and Red Danish.

HOLSTEIN. The Holstein is the largest and most widely used dairy breed in the world. Originally part of the Freisian breed of Holland, Holsteins were first imported to the New World from England in 1652. A large herd was established in the United States in 1852. Today, this breed dominates the American dairy industry.

A Holstein may be distinguished from other dairy breeds by its clearly defined black and white markings, long body, broad hips, and horns that slant forward with an upward curve. Cows weigh an average of about 1500 pounds. Bulls weigh 2200 pounds or more.

Holsteins produce more milk than any other breed. The average yield is about 15,000 pounds per year. Because this breed adapts easily to different climates, has a tremendous resistance to diseases, and a long herd life, it is very popular with farmers.

*A herd of Holsteins resting in a meadow*

JERSEY. Another important breed, the Jersey, derives its name from the island in the English Channel where it was developed. The Jersey was first imported to the United States in 1850. It is the smallest of the major dairy breeds, with cows weighing only about 1000 pounds, bulls about 1500. This breed is usually fawn in coloring, but brown, gray, black, and spotted Jerseys are not uncommon. Some have white markings on the face and hoofs. Their horns typically point forward.

Jersey cows produce milk that is high in butterfat. However, they have the lowest yield (9700 pounds per year) of any of the major breeds. Nevertheless, they continue to be very popular with farmers because they respond well to adverse grazing conditions. They are also easy to manage.

GUERNSEY. Another island in the English Channel, Guernsey, lends its name to a dairy breed. Guernsey cattle first entered the United States in 1830. Today they are found in almost every state.

The Guernsey is a well-proportioned dairy animal. Its head is long with tapered horns that curve forward and upward. This breed is usually fawn with white markings—the concentration of each color varies from one animal to another. Cows and bulls average 1100 and 1700 pounds respectively.

Known as efficient milk producers (about 10,500 pounds per cow per year), Guernseys are docile creatures and can withstand extreme temperatures.

BROWN SWISS. As the name indicates, Brown Swiss cattle are from Switzerland. They were first imported to the United States in 1869. This breed may be distinguished from other dairy breeds by its deep brown coloring and black hoofs. The horns are usually white with black tips. Cows weigh about 1400 pounds, bulls about 2000.

The Brown Swiss cow is widely used by owners of small farms

*A purebred Jersey bull*

*Purebred Jersey cows such as these are an important resource on the island of Jersey in the English Channel.*

*Note the tufted tail and udder of this Brown Swiss cow. Her name is Century Acres Liz C.*

because it is able to produce a large quantity of milk (about 12,000 pounds annually), and it does not require a great deal of care or attention.

AYRSHIRE. Ayrshire cattle were first brought to the United States from Scotland in 1860. A distinctively lean but sturdy strain, the Ayrshire has horns that curve upward and outward. The typical coloring of this breed is red or brown with white splotches. Cows weigh about 1200 pounds. The average weight for bulls is close to 1800 pounds. Ayrshire cows produce more milk than Guernseys, but less than Holsteins or Brown Swiss. However, the milk of the Ayrshire cow is high in butterfat.

RED DANISH. A relatively new dairy breed is the Red Danish,

which was developed in Denmark in 1878 and imported into the United States in 1935. The characteristic red coloring is found in every member of this breed with little variation. A mature Red Danish cow weighs about 1300 pounds and will produce about 9500 pounds of milk per year. Bulls weigh about 1800 pounds.

## Beef Cattle Breeds

All bovines have the ability to convert the food they eat into large quantities of edible flesh, some more than others. Over the centuries, certain strains have been raised exclusively for their meat. These are known as the beef cattle breeds.

The development of the modern beef cattle breeds began in the latter part of the eighteenth century, primarily in Europe and the British Isles. Certain beef types have become more popular than others. To list them all in a volume this size would not be possible. Here are the major beef breeds found in the United States, Canada, and Great Britain:

HEREFORD. These white-faced cattle originated near Hereford-

*A polled Hereford bull*

*An Angus steer*

shire, England, where they were used primarily as beasts of burden and slaughtered for meat when they could no longer work. According to the American Hereford Association, a man named Benjamin Tomkins was the first person to breed the Hereford for its beef characteristics. He did this in 1742. American statesman Henry Clay is generally credited with being the first to import the Hereford to the United States. By 1817, it was discovered that the Hereford could easily withstand the rugged climate of America's western rangelands.

Physically, the Hereford is tall and muscular with a short neck, broad head, and straight legs. Its hide is usually heavy, with a thick, curly coat of hair. The typical coloring is reddish-brown with white markings. Many Herefords have horns, but a polled variety is gaining popularity in the United States. The average weight is 1900 pounds for bulls and 1500 pounds for cows.

ANGUS. The Angus (also called Aberdeen-Angus or Black Angus) originated in Scotland and was imported to the United States in 1873. Initially the breed was used solely on farms in the Midwestern states, but now it is found throughout the country.

Short legs, a stumpy neck, and a bulky body distinguish the Angus from other beef breeds. Cows and bulls generally average about one hundred pounds more than Herefords of the same age and sex. The Angus is considered to be an excellent beef animal because of its wide loin, long rump, and bulging quarters. Angus cattle are usually polled.

A related strain, the Red Angus, is similar except for its red coat. Since red absorbs less of the sun's heat, this breed is often raised in hot climates.

SHORTHORNS: The Shorthorn originated in England during the late eighteenth century. It was bred first as a dual-purpose animal. Today there are two types of Shorthorns—a beef variety, which is raised solely for its meat, and the so-called Milking Shorthorn, which is used for both milk and beef.

Beef Shorthorns are very compact and heavily muscled. Cows weigh about 1800 pounds with males averaging close to 2500

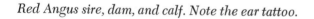

*Red Angus sire, dam, and calf. Note the ear tattoo.*

*A Shorthorn cow with her calf*

pounds. A Shorthorn may be red, white, or roan. The latter is a color combination which results when a red and a white animal mate. As the name indicates, the horns are short, but a polled strain is now seen on many ranches.

Shorthorns were initially imported into America in 1783. Between 1820 and 1850, large herds were established in the United States. Today they are popular with cattle raisers because of their good temperament and manageability.

CHAROLAIS. No discussion of beef cattle would be complete without mentioning the Charolais, a French breed imported to the United States from Mexico in the 1930's. This breed is usually white or cream colored. Charolais are good grazers and can withstand extreme heat or bitter cold.

BRAHMAN. In the early 1900's, the Brahman was developed in the United States from four types of *Bos indicus*. Physically, the Brahman is very similar to its progenitor in appearance.

44

Contrary to popular belief, purebred American Brahmans such as this one are intelligent animals bred for gentleness. They respond affectionately to proper treatment.

*A Santa Gertrudis bull, the first truly American breed*

The typical coloring is light or medium gray. However, red or black Brahmans are not uncommon. In the southern part of the United States, Brahmans are very popular because they can graze all day in hot sun and are not bothered by insects. The Brahman is the basis for all of the beef breeds that have been developed in the United States.

SANTA GERTRUDIS. The first truly American cattle breed is the Santa Gertrudis, named after a region of the King Ranch in Texas where it was developed in 1918. A cross between the Brahman and the Shorthorn, the Santa Gertrudis is typically deep cherry red in coloring. Its hide is loose, and in warm climates it has short, straight hair. In colder areas, it grows a longer coat. Prominent characteristics also include neck folds and drooping ears. Some Santa Gertrudis cattle are polled. Most have horns that slant downward.

OTHER BEEF BREEDS. By crossing the Hereford and the Brahman, cattle raisers have produced the Braford. Herds of Brangus have been established by breeding Brahmans with Angus cattle. Several generations of breeding were required to develop the Beefmaster, which is a combination of Hereford, Shorthorn, and Brahman cattle.

## Dual-Purpose Cattle Breeds

Dual-purpose breeds have good beef qualities and are able to produce reasonably large quantities of milk. Most dual-purpose cattle are used in Russia, South America, and Europe. There are

*Crossbreeding Brahmans with Angus cattle produced the Brangus.*

A Grand National Champion Milking Shorthorn

A Red Poll cow

several types of dual-purpose cattle found in the United States.

As noted earlier, the Milking Shorthorn is similar to the Beef Shorthorn. Both have the short horns, but the dual-purpose variety has the long, angular features of dairy cattle. Breeders of the Milking Shorthorn have developed cows that yield large quantities of milk (about 11,000 pounds per year) and calves that produce excellent veal.

The Red Poll was imported to the United States from England in 1873. As the name indicates, this strain is red in color, and it has no horns. It may be distinguished from other breeds by its white tail switch, long legs, and the absence of the rear udder depth found in dairy cattle. Nevertheless, the Red Poll produces more milk than beef breeds. Its carcass yields a good quantity of lean beef which is low in fat. Red Polls are particularly popular with farmers who want to run a dairy and also sell steers for beef.

# 5 The Cattle Industry

*Breeding*

Every science has its own vocabulary—words that have a special meaning to the people involved in a particular field of study. Breeding, the science of improving livestock by controlling reproduction, is no exception. Here is a brief sampling of the terms used by people engaged in cattle breeding.

The owner of a calf's parents at the time they were mated is called a breeder. "Character" (or breed character) refers to the physical features that distinguish one breed from another. A bred cow is a pregnant female. If she turns out to be a good mother, she is called a "broody." A cow used primarily for breeding purposes is known as a dam. Her male counterpart is a sire.

Furthermore, the science of breeding is regarded by many as an art. The ability to identify the best animals for breeding is a talent that takes years to develop. Some say it is a gift. In many cases, this ability is passed down within a family from one generation to the next.

People who engage in cattle breeding as an occupation usually specialize in a certain type. Some raise only bulls—others only cows, cows and bulls, or cows and calves. A person who raises

*Hereford bulls receive special dietary supplements to improve their efficiency as breeding stock.*

just cows, for example, tries to improve his herd so that ranchers or farmers will buy from him to build up their own stock. There is no "best" breed. Buyers usually purchase cattle of the strain that suits their own individual needs.

Many breeders specialize in a type of breeding. Some use straightbreeding—raising animals of one specific breed. There are two methods of straightbreeding—outbreeding (mating cattle that have no common ancestors) and inbreeding (mating related cattle). Crossbreeding—the mating of different breeds —is also widely practiced. This method of breeding has been successfully used to improve beef cattle. However, crossbreeding is both expensive and risky. The breeder never knows what the results will be.

*This rancher drives his car into the pasture to check on the condition of his pregnant cows. His daughters, left, like to feed broken ears of corn to the tame animals.*

The modern cattle breeder controls the number of calves a particular cow will bear during her lifetime. This is accomplished simply by not mating the cow. The cow's gestation period is about 280 days. The ideal season for calves to be born is in the spring, when the best fodder is available. Thus, cows are usually bred the summer before a calf is desired.

Scientific techniques are being used to improve the quantity and quality of breeding stock. Computers keep accurate records of milk or meat production, as well as the growth rate of heifers and steers. These statistics aid the breeder in selecting the best breeding stock. Scientists have also developed a way to impregnate cows artificially. This process enables an oustanding bull to produce thousands more offspring than he would by natural means. Additionally, the owner of a small ranch or farm can take advantage of a superior breeding animal that would otherwise be too expensive to transport to his cows.

Some cows give birth prematurely—as many as ten days earlier than expected. In most breeding operations, bred cows are separated from the rest of the herd shortly before the calving date. On the open range, the mother-to-be instinctively goes to a quiet place to have her calf.

Most births are uncomplicated. Occasionally, however, something goes wrong. If the calf's tail emerges first, the animal may suffocate. Or if the mother becomes tired, she may be unable to complete the delivery. The breeder or a veterinarian usually observes the birth and helps if there are any problems.

If the calf survives the birth but the mother dies, the newborn is given to a broody who has recently given birth. However,

*A future farmer takes water to a cow that was partially paralyzed in giving birth to her calf. As therapy, the cow will be helped to its feet and encouraged to walk, which will speed its recovery.*

*Cows, like this Angus, are affectionate mothers.*

not every cow will accept a calf that is not her own. If she rejects it, the breeder cares for the calf until it can take care of itself.

Normally, a newborn calf receives a milky substance called colostrum from its mother for the first three days of its life. It is given whole milk from its mother or from a bottle for about three months. Then it is weaned and fed grain or other fodder.

## Dairying

The answer to the old Albanian riddle "How does a brown cow eat green grass and give white milk?" has long been forgotten—and rightfully so. Cows need not be brown to give milk, nor do they eat only green grass. The diet of the cow also consists of hay, oilseed meal, grain, corn, and vitamin supplements.

The question remains, then, how do cows give milk? To simplify, milk production involves the digestive system, bloodstream, and udder. Nutrients from digested food are brought by the bloodstream to the udder, where glands convert them into milk.

*Elsie the Cow, mascot and trademark of Borden, Inc., a major producer of dairy products. Elsie's home is in Columbus, Ohio. This photograph was taken when Elsie visited Van Saun Park in Paramus, New Jersey.*

Young Holsteins, in stanchions, feeding on hay

Today, most cows are milked by machines.

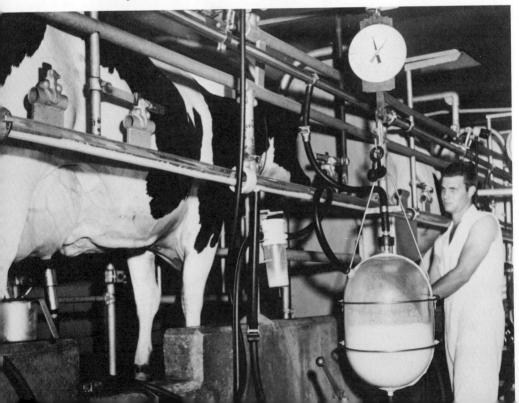

A cow reaches its peak as a milk producer about two months after it bears a calf. From then on, the quantity of milk produced declines. On most dairy farms, a cow is milked until about two months before she is due to have her next calf.

A cow can be milked by machine or by hand. Sanitary conditions must be maintained before and after milking to prevent contamination of the milk.

*Cleverlands Nances Judy Jean, a National Class Leader for four consecutive years at age six, produced 33,380 pounds of milk in one year. She's a purebred Guernsey.*

Before a cow is milked, the udder is washed and massaged. Once the milk has been withdrawn (usually by squeezing the udder by hand or machine), it is carried in cans or by pipeline to a large tank in a refrigerated area called the milk room. From there, it is usually shipped by truck or train to a processing plant.

Successful milking depends largely on being able to get the

cow to "let down" her milk. This can be done by establishing a daily routine—feeding, exercising, and milking her at the same time each day. Incidentally, a cow can recognize a certain sound (such as the rattling of pails or the hum of a milking machine's engine) as a signal to let down her milk.

Most of the milk consumed in America is produced in a region called the Dairy Belt—an area that runs from New England to eastern Iowa.

### Ranching

The term "ranching" is commonly applied to the care and management of horses, sheep, and cattle. Most American beef cattle are located in the western states, where pasture is plentiful, and in the north central region, where large quantities of grain are produced. These areas are often referred to as "cattle country."

Cattle ranching has changed considerably since the era of the

*A rancher moves cows and calves out of a corral after inspecting them for health and nutrition problems.*

*A cowboy rides herd on his purebred beef cattle. Scenes like this on a ranch in Utah are common in the western United States, which is often referred to as "cattle country."*

Old West. While the independent rancher is still an important part of the American cattle industry, most ranches are owned and operated by corporations. The person who manages a ranch usually has a degree in breeding science or livestock care and has taken courses in how to obtain the best price for a herd.

For the most part, cattle are driven to market in trucks and railroad cars. However, the high cost of fuel has forced many ranchers to move their herds on foot, just as their ancestors did. The modern cattle trails are often paved with concrete or macadam and, instead of fighting off rustlers, cowboys must contend with tourists who pollute waterholes and litter the countryside.

A great deal of the work done on today's ranches is performed by machines. Trucks bring fodder to the herd. Helicopters are used on large ranches to count the number of head in a particular area, locate injured stock, or drive the herd in from the range. Automatic sprinklers provide water during dry spells.

Today's cowboy is a person who is hired to watch over a herd

*The wagon of hay in the foreground is used to lure cattle into the corral. As the cattle move in, cowboys use gates to separate cows from calves.*

while it is out on the range. He still spends his days in the saddle, pulling dogies out of bramble bushes or chasing coyotes, but at night he sleeps in a trailer.

For most of the year, beef cattle are allowed to roam freely. In some areas, the winter is the hardest season, since the herd must be given extra fodder and sometimes shelter. Open sheds or barns serve this purpose.

The spring roundup is an annual event on most ranches. At this time, the calf crop (calves born in any one year) is harvested. When the herd is brought in from the range, the calves are cut out for branding and tattooing.

60

*Separating calves from their mothers is an important early step in the branding operation. The calves are in the foreground.*

*These ranchers spent over three hours to see that 325 calves were innoculated against disease and branded. The males are also castrated, and one ear of each female is notched with a knife to make her easier to identify on the range as she gets older.*

Branding is now accomplished more easily. The calf is herded into a V-shaped metal trough called a squeeze chute. Here a clamp is placed around its neck. This holds the calf in place while the squeeze chute is tilted on its side to form a branding table.

The use of a hot iron is still the preferred means of branding. Other methods have been developed. Cold branding involves the use of a copper rod that has been dipped in liquid nitrogen. The extremely cold temperature ($-90°$ Fahrenheit) kills the pigment cells on the hide, leaving a white brand. However, cold branding does not work well on animals that have white hair. Some ranchers use caustic acid to burn in a brand but this is considered to be even more painful than the use of a hot iron.

After a calf has been branded, its ears are tattooed with a number for identification. An accountant notes this number so that the rancher will have a record of his inventory. Calves to be sold as steers are castrated. Horns are removed, if present, and all are vaccinated.

Several diseases commonly plague bovines. The three most serious ones are blackleg, malignant edema, and brucellosis. All are fatal but can be prevented by innoculations. Incidentally, calves are usually dipped in a liquid insecticide to kill ticks and screwflies which burrow into their hides and cause illness.

Many ranchers maintain two herds—a breeding herd, which is used for producing more stock, and a grade herd (the larger of the two), which consists of calves that are to be sold for beef. Some ranchers fatten their grade herd or sell it to others who specialize in preparing stock for slaughter. The fattened cattle (called feeders) are graded by the owner from fancy to inferior, based on their expected ability to gain weight and thus yield a large quantity of beef. Cattle destined for slaughter are fed considerable amounts of hay, silage, and corn for about one hundred days. Most reach the market at twenty months of age, weighing about one thousand pounds. At this point, they are considered to be "finished" cattle and are sold at auction to meat packing companies.

# 6 A Very Valuable Animal

*Milk*

While all female mammals nurture their young with milk, the cow is by far the greatest producer. In fact, when people speak of milk, they are usually referring to the milk of the cow.

Except for water and possibly wine, milk is the most widely consumed liquid in the world. Its overwhelming popularity is no mystery since milk is an important source of nutrition. It contains proteins, mineral salts, and lactose (milk sugar). Milk also has a pleasing taste. When served cold, it quenches the thirst. Warmed milk is said to soothe the stomach or help a person fall asleep.

Many popular products are obtained from milk. In its raw form, this fluid contains globules of a substance called butterfat. When butterfat rises to the top of a container of milk, it forms cream, which can be churned into butter or whipped for dessert. By coagulating the protein in milk, cheese can be made.

Removing some of the water produces evaporated milk, which is often used in feeding small children. Adding sugar to evaporated milk yields condensed milk, which is required for many dessert recipes. Milk sugar is used as a food supplement in the diet of infants. Casein, a protein in milk, can be made into glue or hardened to form billiard balls, combs, buttons, or piano

*This Holstein calf is only a few days old.*

keys. Milk can also be processed into buttermilk, whey, curds, malted milk, powdered milk, and ice cream.

## Meat

Meat is the term applied to the edible portion of wild or domestic mammals. Cattle, after they are slaughtered, provide two types of meat—beef and veal. Beef, which is deep cherry red in

*These cattle are being loaded into railroad cars for shipment to the packing plant.*

color, is obtained from mature cattle. Veal is the flesh of calves slaughtered when they are six to twelve weeks old and is usually grayish-pink in color. The most desirable veal is obtained from milk-fed calves that are killed when they are six weeks old. A calf bred for veal is usually taken away from its mother at birth, placed in a stall, and fattened until it is ready to be slaughtered.

Meat, especially beef and veal, is a highly nutritious food which contains proteins, vitamins, and iron. The United States is the largest producer of beef in the world, with about sixty million head of cattle slaughtered each year. Canada, Argentina, and Australia are also large beef producers.

The beef industry has a language of its own. After an animal is slaughtered, its hide is removed. What remains is called a carcass, which is split into halves called sides. Cutting a side of beef in half produces two sections—the hindquarter and the forequarter. Beef is shipped to the butcher in quarters, while veal is transported as a whole carcass.

The butcher divides the quarters into what are called retail cuts—the ones the consumer buys in the store. The names given to these vary from one part of the country to another.

Shoulder cuts are called chuck and can be made into pot roast and chuck steak. Rib cuts produce rib steaks and roasts. The rump is the upper part of the hind leg. The loin is cut into porterhouse, T-bone, and sirloin steaks. Flank is a boneless cut from the underside of the animal. The plate is a long, thin piece located in the forequarter in front of the flank. The upper part of the foreleg is called the shank. Beef can be processed into frankfurters, bologna, and sausage, or dried to produce jerky. The liver, brains, tongue, kidneys, and intestines are sold as variety meats.

All beef produced in the United States is inspected by the Department of Agriculture, which grades it according to quality. However, the protein, vitamin, and mineral content is the same for all beef, regardless of grade.

## Leather

There's an old fable from the Middle East that tells about a conversation between an ox and a horse. The two animals are spending an afternoon boasting about their respective accomplishments. The horse proudly states, "If it wasn't for *me*, Man wouldn't be able to go anywhere." The ox nods, almost in agreement. "That may be true," he says, "but if it wasn't for *me* you wouldn't be so successful." The horse is puzzled. "Why is that?" he asks. The cow replies, "*I* provide the leather to make your saddle and harness. Without *me*, Man wouldn't be able to use *you*."

The story probably was meant to illustrate Man's dependence on cattle for a basic raw material—leather. As with milk and meat, cattle are the major suppliers.

There are several kinds of leather made from cattle hides. These range from supple and lightweight types to strong and tough types. Calfskins supply most of the softer leather for wallets, luggage, and shoe uppers. The tough leathers (produced by older animals) provide the raw material for shoe soles, harnesses, belts, saddles, and a variety of other goods.

## Other Uses

Man has found a use for practically every product obtained from cattle. Cow dung (commonly called manure) is packaged and sold as fertilizer. Farmers spread manure on their fields to increase crop yield. In many parts of the world, hardened dung is burned as heating and cooking fuel.

Traditionally, the horns of the ox are a symbol of fertility. Since horns are hollow and almost funnel-shaped, they have been used for carrying small quantities of gunpowder. The shape is particularly useful when pouring the powder into

*Note the ears on this Angus calf. "Camel's hair" brushes are actually made from the hair on a calf's ear.*

muzzle-loading weapons. Additionally, the horn has served as a signaling device for centuries. By removing the tip and blowing into the narrow end, a sound can be produced. Incidentally, brass musical instruments such as trumpets and trombones are played in a similar fashion and thus are often referred to as horns.

The fat of this animal is used in grease, candy, gelatin, and margarine. Plywood adhesive, pharmaceuticals, and textile dyes are made from the blood. The bones of these animals are ground up for use in glue, neat's-foot oil, fertilizer, and livestock feed. By the way, "camel's hair" brushes are really made from the hair on a steer's ear!

# 7 Rodeos and Shows

Rodeos are contests based on the skills developed by cowboys during the cattle raising era of the Old West. These contests originated informally during the 1840's. In those days, it was common for cowboys to celebrate the end of a roundup by wagering on their individual skills.

The first formal rodeo was held in Pecos, Texas, on July 4, 1883. It was a huge success. As time passed, rodeos became large extravaganzas that drew contestants and spectators from all over the country.

Several rodeos have achieved national fame. Frontier Days of Cheyenne, Wyoming, originated in 1897 and is considered to be the oldest annual show of its kind. The Calgary Stampede of Alberta, Canada, has been held every year since 1919. Indoor rodeos in arenas such as New York's Madison Square Garden are also very popular.

The scene at a rodeo is like returning to the era of the Old West. Cowboys and cowgirls dress in ten-gallon hats, denim jeans, chaps, and boots. While some private wagering still takes place, contestants compete primarily for prize money and fame. A successful rodeo participant can earn the same adulation as a professional football player or a movie star.

Cattle are used for three main rodeo events—bull riding,

steer wrestling, and calf roping. These events require a great deal of skill and concentration.

In the calf-roping event, a calf and a cowboy on horseback are released from the same chute. The calf receives a slight head start. The rider must pursue the calf, lasso it, and drag it to the ground. Next, he must tie three of its legs together with a small rope. The contestant who accomplishes this in the fastest time is the winner.

The object of bull riding is to stay on the back of a bucking bovine for at least eight seconds. This is no simple task, since the bulls used for this event are ornery critters that won't let *anyone* ride them. To help him stay on top of the bull, the rider uses a braided rope which is wrapped once around the beast's stomach. The rider holds one end in his strong hand and may not touch the bull with his free hand at any time. If he does, he is disqualified.

*Calf roping pits skilled rider and horse against a calf and the clock.*

*Bull riding takes strength and skill. Note the rodeo "clown," right.*

The rodeo announcer shouts "Go!" and the fans yell "Ride 'em, cowboy!" The door to the chute holding the bull and its rider is opened. As the bull comes charging out, the rider tries his best to hang on.

Many contestants feel that bull riding is the most dangerous rodeo event. If the rider is thrown by the bull or if he manages to stay on for eight seconds and then jumps off, he must release the rope immediately. If he doesn't, he could become tangled in it and dragged along the ground. Once the rider is on the ground, he is not out of danger. The bull could charge. If that happens, "clowns" (people dressed in brightly colored costumes) try to "shoo" the bull away, risking their own lives in the process.

In steer wrestling (which is also called bulldogging), the contestant, who is mounted on horseback, rides after a steer that has been released from a separate chute. Once he overtakes the animal, the contestant leaps from his saddle. He tries to grab the steer's horns, simultaneously digging his heels into the ground. As he does this, he twists the steer's head, bringing them both to an abrupt halt. Next, he must force the animal to the ground so that all four of its legs are pointing in the same direction. The contestant who accomplishes this in the shortest amount of time is the winner.

There is always the risk, in steer wrestling, that the contestant may miscalculate his leap and land on the steer's horns. If he jumps too soon, he could be trampled to death. More often than not, however, the rodeo contestant's skill enables him to make his leap correctly.

Some of the events at rodeos are held purely for fun and for entertainment. Children compete in the calf scramble by trying to drag a reluctant calf into the center of a ring. The first youngster to succeed is the winner. Steer decorating, cow milking, and ox handling are other popular events.

There are people who argue that some of the rodeo events mentioned here cause injuries to the animals and are, therefore, inhumane and cruel. This may be true. Nevertheless, rodeos will continue to be an American tradition.

Cattle shows are also an American tradition. Each year, thousands of people showcase their cattle at county and state fairs throughout the United States and Canada. Breed associations also sponsor cattle shows.

Producing a show champion begins with the choice of a calf and includes a careful program of training, feeding, and groom-

*The Calgary Stampede and Exhibition, Alberta, Canada, includes a cattle show.*

ing. If the calf is chosen when it is young, it can be trained without much difficulty. Before a calf can be exhibited, it must be taught to pose and lead. Grooming is also important. Many exhibitors wash, trim, brush, and curl the animal's coat on the morning of the show. The horns and hoofs are oiled so that they will be shiny.

"Show" cattle are usually purebred dairy or beef animals. There are many categories in which cattle may be exhibited. They generally compete against each other according to breed, sex, age, and weight class. The actual competition takes place in a ring. Judging is often based on the animal's breed character,

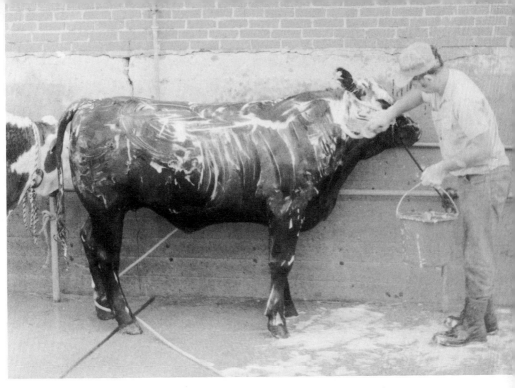

A steer receives a scrubdown from its owner before a show.

A vacuum cleaner is used to clean this steer on show day.

*A Jersey island champion cow, and a prize-winning bull, facing page*

conformation, and physical appearance. The exhibitor's ability to control the animal is also evaluated.

Prize money, blue ribbons, or trophies are awarded to the grand champion, first, second, and third place winners, plus honorable mention. A grand champion steer, bull, cow, or calf can be sold at auction for about three times as much as a non-winner. Most exhibitors, however, do not compete solely for financial gain. To many, the recognition received for showing a champion is a source of family pride. Participating in a show is also an excellent opportunity for future cattle raisers to learn more about these valuable animals and all their wonders.

Right: *Holstein cows wait to be judged at a cattle show. Breed associations such as the Holstein Association of America often sponsor these shows.*

# Index

American Hereford Association, 42
Angus, 43, 47
Apis, 17
*Augsburg Aurochs Picture*, 20
Aurochs, 10, 11
Austin, Stephen F., 25
Ayrshire, 37, 40

Babe, 20
Baltic Finns, 16
Banya, 19
Beefmaster, 47
Beef Shorthorn, 43–44, 47
*Book of the Dun Cow*, 20
*Bos indicus*, 9–11
*Bos longifrons*, 11
*Bos primigenius*, 10, 11
*Bos primigenius nomadicus*, 11
*Bos taurus*, 9–10
Bovidae, 9–10
Bovinae, 9
Braford, 47
Brahman, 44–45, 47
Brangus, 47
Breeding, 50–54
  crossbreeding, 51
  inbreeding, 51
  outbreeding, 51
  straightbreeding, 51
Breeds, 35–49

beef, 35, 41–49
dairy, 35–41
dual purpose, 35, 47, 49
Bres, 18
Brown Swiss, 37–38, 40
Bull baiting, 16
Bull fighting, 16
Bull riding, 70–73
Bunyan, Paul, 20
Burgess, Gelett, 20

Calf roping, 71
Calgary Stampede, 70
Cattle
  in art, 14, 20–21
  as beasts of burden, 7
  Castillian, 24
  domestication of, 11–12
  European, 9
  history of, 9–13
  Indian, 9
  in language, 7–8, 21
  leather, 7
  in literature, 20
  in lore, 14, 16–18
  meat, 7
  milk, 7
  Moorish, 24
  origin of the word, 7
  oxen, 7, 12

in religion, 14, 16–20
  wild, 7
*Cattle Raid of Cooley*, 18
Celtic shorthorns, 11
Celts, 18
Charolais, 44
Clay, Henry, 42
Columbus, Christopher, 24
Coronado, Francisco Vasquez de, 24
Cowtowns, 31

Dairy Belt, 57
Dairying, 54–58
Drago, Harry Sinclair, 24, 26

East African Cattle Area, 19
Egyptians, 12, 16, 17, 20
*Eotragus*, 10
Exodus, Book of, 20

Finns, 18
Frontier Days, 70

Glidden, Joseph F., 33
Golden Calf, 20
Gufittar, 18

Hereford, 41–42, 47
Hindus, 19
Holstein, 37

Isis, 17

Jason, 17

Kadmos, 18

Lear, Edward, 20
Leather, 68
Longhorn, 25–27, 34

Maahiset, 18
Madison Square Garden, 70
Masai, 19

Maverick, Samuel, 28
Meat, 65–67
Mesopotamians, 17
Milk, 64–65
Milking Shorthorn, 49
Mongolians, 18
Moses, 20

Nekhbet, 17
Norsemen, 24

*Odyssey*, 17
Old Testament, 20
Old West, 24
Orion, 18
Osiris, 17

Pamplona, Spain, 16

Ranching, 57–63
Red Angus, 47
Red Danish, 37, 40–41
Red Poll, 49
Rodeos, 70–73
Romans, 13
Root, E. M., 20
"Running of the Bulls," 16

Santa Gertrudis, 46
Scandinavian Lapps, 18
Shows, 73–76
Siberians, 18
Sothis-Sirius, 17
Steer wrestling, 73

Taurus, 21
Texas, University of, 22
Theseus, 14
Tomkins, Benjamin, 42
Tuath Dé Danann, 18
Tutsi, 19

Zeus, 16

## About the Author

Vincent Scuro, whose first book, *Presenting the Marching Band*, was published by Dodd, Mead & Company, has an avid interest in wildlife preservation and photography. An accomplished musician on the trumpet, guitar, and bass guitar, he played baseball in college, and has worked as a newspaper correspondent. His photographic work has appeared in books and magazines.

Now a consultant with a computer-based education company, the author has taught social studies and coached volleyball in high school. He lives in Hillsdale, New Jersey.

Vincent Scuro is the co-author of *Wonders of Goats, Wonders of Elephants, Wonders of Donkeys*, and *Wonders of the Bison World*.